I am going to Granddad's ranch. Its name is Tumbleweed Ranch. I have been there lots of times!

Sometimes I see a tumbleweed tumble. It looks like a little bush rolling on the grass.

Today we get up at sunrise. We put maple butter on warm pancakes. Yum!

Granddad goes to the stable. I help him pull a cart full of hay for his horse. Her name is Little Red. But she is big!

Granddad puts on her saddle. We want to go for a horseback ride.

I sit in the saddle with Granddad.
We ride up a hillside.

We see cattle on the hillside. One of them has a bell that goes jingle-jangle.

Back at the ranch I get an apple for Little Red. I bet her little nod says "Good!"